YOUR KNOWLEDGE HAS VALUE

- We will publish your bachelor's and master's thesis, essays and papers

- Your own eBook and book - sold worldwide in all relevant shops

- Earn money with each sale

Upload your text at www.GRIN.com and publish for free

Bibliographic information published by the German National Library:

The German National Library lists this publication in the National Bibliography; detailed bibliographic data are available on the Internet at http://dnb.dnb.de .

Imprint:

Copyright © 2009 GRIN Verlag, Open Publishing GmbH
Print and binding: Books on Demand GmbH, Norderstedt Germany
ISBN: 9783640598861

This book at GRIN:

http://www.grin.com/en/e-book/149394/lighter-than-air-concepts

Sascha Hissler

Lighter Than Air Concepts

LTA-crafts. An overview

GRIN Publishing

GRIN - Your knowledge has value

Since its foundation in 1998, GRIN has specialized in publishing academic texts by students, college teachers and other academics as e-book and printed book. The website www.grin.com is an ideal platform for presenting term papers, final papers, scientific essays, dissertations and specialist books.

Visit us on the internet:

http://www.grin.com/

http://www.facebook.com/grincom

http://www.twitter.com/grin_com

AVIMA

2008

Third Semester

MANUFACTURERS MANAGEMENT

LIGHTER THAN AIR CONCEPTS

November 2009

SASCHA HISSLER

Contents

1. Introduction

When thinking about lighter-than-air concepts many people have the fascinating pictures of large Zeppelin airships and the Hindenburg disaster of 1937 in mind.

At first view, making use of buoyancy seems to be an old-fashioned approach to human flight that does not comprise significant technical challenges. But even today, there is a small fleet of technically relatively simple and small-sized airships operating over the world - mainly for the purpose of advertisement and commercial leisure flights. The most famous and oldest lighter-than-air concept is the traditional hot air balloon - available for passenger rides and fun. On the other side, a growing global interest and quite ambitious activity in developing large high-tech airships for specific promising applications (e.g. cargo lifting) can be observed in the field of LTA-crafts.

But what exactly is an lighter-than-air craft? An aircraft is anything that flies. A lighter-than-air craft is any vehicle that flies because it is lighter than air, like balloons, airships and aerostats[1]. What makes it lighter than air is the fact that it uses a lifting gas (e.g. helium, hydrogen, hot air) in order to be lighter than the surrounding air and by this making use of buoyancy. Any gas or gas mixture that has a lower density than air is "lighter than air" and therefore suitable as a lifting gas. Under standard conditions (0°C, 1bar) air has a density of 1.275 kg/m^3. To see the difference of lifting power, there is the rule of thumb that 1m^3 of hydrogen lifts 1.1 kilogram, 1m^3 of helium lifts 1 kg and 1m^3 of hot air lifts only 300 grams.[2]

2. Hot Air Balloons

Hot air balloons are the oldest successful human-carrying flight technology, dating back to the Montgolfier brothers' invention in France in 1783. The first flight carrying humans was made on November 21, 1783, in Paris and stayed in the air for a grand total of 20 minutes.[3]

---

[1] Derived from the greek words "aer" and "statos", which means standing or staying in the air.

[2] http://www.peacesoftware.de/einigewerte/luft.html

[3] http://www.eballoon.org/history/history-of-ballooning.html

A hot air balloon consists of three parts: the basket, the burner and the envelope. Latter is the huge bag which is capable of containing hot air. Suspended beneath is the (wicker-) basket, also called gondola or capsule, which carries a source of heat capable of producing a sufficient temperature gradient between the air inside the envelope and the surrounding air mass to give enough lift to keep the balloon and its passengers aloft. Air that has been heated to about 70°C has a lower density than atmospheric air. Its density is approximately 1.02 kg/m^3 and therefore significantly higher than that of hydrogen or helium resulting in less lift. As the balloon derives lift from the hot air which is blown into the hull of the balloon and makes it fly. It is then following the direction of the wind (horizontal movement) and usually has a mechanism to control the vertical movement. A great disadvantage is that heat has to be added to the air constantly because otherwise it cools down and consequently loses lift as density increases with cooling.

Unlike gas balloons, the envelope does not have to be sealed at the bottom since the rising hot air only exerts pressure on the upper hemisphere of the balloon to provide lift.[4] In today's sports balloons the envelope is generally made from nylon fabric and the mouth of the balloon, which is located close to the burner flame, is made from fire resistant material such as Nomex.[5]

A large-scale technical application of hot air for the later described airships is impossible, because high heating costs and the low amount of lift makes it uneconomical.

That is why hot air balloons are used primarily for commercial leisure flights, recreation and as a sightseeing attraction for tourists nowadays. The hot air balloon baskets may vary in size and carry anything between 3 up to 20 passengers.[6] Due to the huge envelope of the balloon, it is often used as advertising space and some of them are even used for sport (records) as well: The record for the longest duration of a hot air balloon flight was set up by the adventurer Steve Fossett, who traveled 20,602 miles (32,963 km) around the world in 14 day, 19 hour and 51 minutes in Summer 2002.[7]

---

[4] http://science.howstuffworks.com/hot-air-balloon1.htm

[5] http://blog.adventureballoons.co.uk/category/the-balloon-envelope

[6] http://www.eballoon.org/balloon/commercial-leisure.html

[7] http://www.stevefossett.com/html/press_archives/balloon_04july02.html

Constitutive parts of a hot-air balloon:

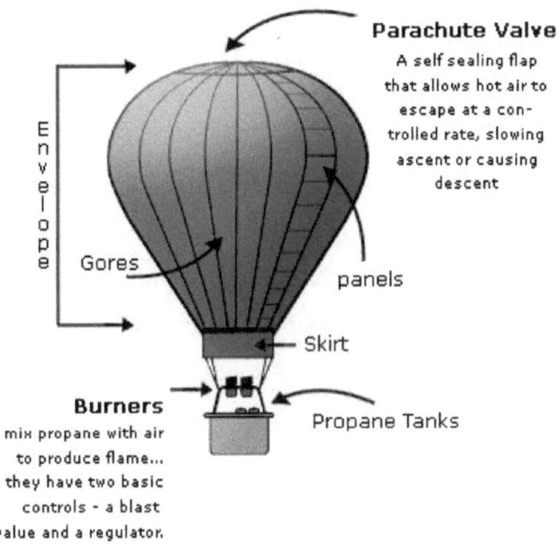

3. Lighter-than-air crafts

In contrast to a balloon, an airship is a powered lighter-than-air craft, which is propelled forward by an engine including some means of controlling the direction, usually with rudders. Because of this advantage these airships are also called "dirigible", which is derived from the french word "diriger" meaning steerable.

We distinguish between three main types of airships: rigid airships, semi-rigid and non-rigid airships, also known as blimps. All three types have four principal parts: a cigar-shaped bag, or balloon, that is filled with a lighter-than-air gas, a gondola that is slung beneath the balloon and holds the crew and passengers and one or more engines that drive propellers as well as horizontal and vertical rudders to steer the airship.

3.1. Rigid Airships

As their name implies, rigid airships are rigidly constructed and keep their shape with an inner latticework of metal supporting a bag containing pockets of lifting gas. The

---

8 http://www.eballoon.org/balloon/how-it-works.html

most famous rigid airships were the Zeppelins and the USS Akron and Macon. The era of the Zeppelins began with the flight of the first *Luftschiff Zeppelin D-LZ-1* over Germany's Lake Constance in July 1900 and ended with the disaster of the *D-LZ-129* (*The Hindenburg*) in May 1937.[9] The most famous of the Zeppelin airships was the *D-LZ-127*, which was launched for the first time in 1928 and was designed for carrying 20 passengers and a crew of 40. This airship set a host of records, including the longest non-stop flight from Germany to Lakehurst, New Jersey, a 6200 mile trip, in 111 hours and 44 minutes. In August 1929, it made the 20,500 mile around the world trip in a total of 12 days in the air.[10] Until the Hindenburg accident in 1937 about 405,000 passengers travelled across the Atlantic by a Zeppelin.[11] During that time, airships like the Zeppelin had an advantage over the airplane on intercontinental flights, but that advantage had vanished by the end of World War II. Airships like these were also used during World War I, but by mid-1917 the airship could no longer survive in a combat situation where the threat was airplanes. During World War II they were still used for anti-submarine patrol flights over the pacific.

[12]

Airship D-LZ 127 "Graf Zeppelin"

[9] http://www.century-of-flight.net/Aviation%20history/up%20to%20WW%201/Dirigibles2.htm

[10] http://www.altfrankfurt.com/Spezial/Zeppelin/GrafZeppelin

[11] http://www.nytimes.com/slideshow/2008/07/06/business/0706-BLIMP_5.html

[12] http://www.altfrankfurt.com/Spezial/Zeppelin/GrafZeppelin/pic/GrafZeppelin3.jpg

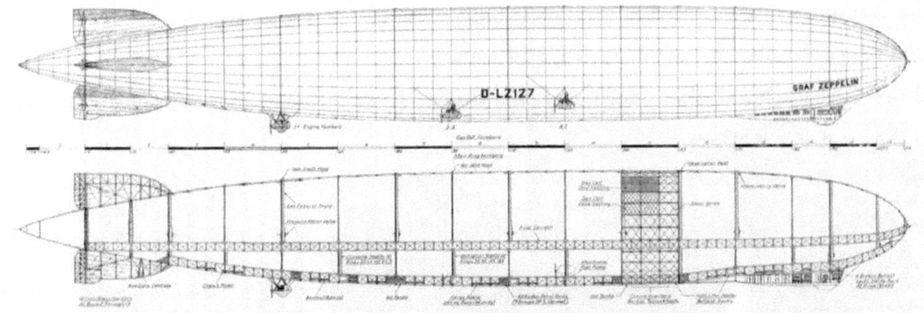

"Graf Zeppelins" profile, showing rings, gas cells, and major elements.[13]

All airships built by "Zeppelin's company, DELAG, from LZ 1 to LZ 129 (Hindenburg) varied in details, but they were all modelled on the principles established by the first one."[14] The rigid structure, traditionally an aluminum alloy, holds up the form of the airship even if no gas is inside. In general, rigid airships are only efficient when longer than 120 Meters (360ft) because a good weight to volume ratio is (or was) only achievable for large airships. For a small airship the solid frame would have been too heavy. The capital disadvantage of the zeppelin airships was that they were "too slow and explosive a target in wartime (over 40 were shot down over London) and too fragile to withstand bad weather."[15]

DIRIGIBLE (airship)

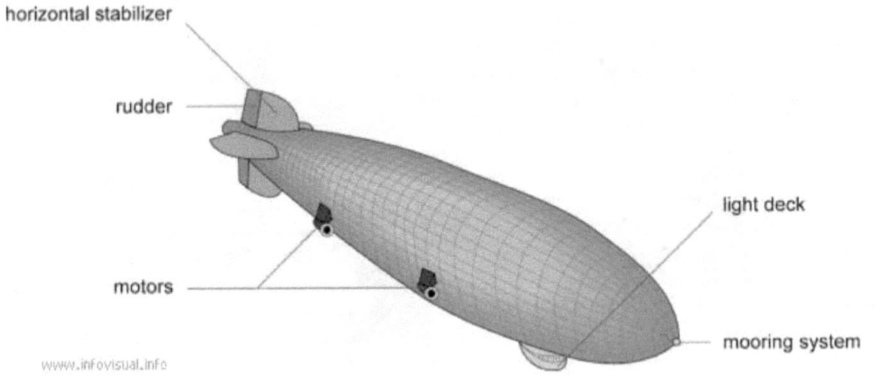

---

[13] http://www.airships.net/lz127-graf-zeppelin/design-technology

[14] http://www.century-of-flight.net/Aviation%20history/up%20to%20WW%201/Dirigibles2.htm

[15] http://www.century-of-flight.net/Aviation%20history/coming%20of%20age/zeppelin.htm

[16] http://www.infovisual.info/05/093_en.html

The balloon of a dirigible is equipped with engines for propulsion and a horizontal stabilizer including a rudder at the end used to steer the vehicle. The crew is located at the light deck and the mooring system is used to immobilize and "park" the airship at the mooring mast:

Airship connected to the mooring mast.

3.2 Semi-rigid Airships

An intermediate version of a rigid and the later described non-rigid airship is the semi-rigid airship, which were more popular earlier last century.[18] They usually comprise a rigid lower keel construction and a pressurized envelope above that. The rigid keel can be attached directly to the envelope or hung underneath it. One of the most famous representative of that type was the Italia, the airship which General Umberto Nobile used on his attempt to reach the North Pole.[19]

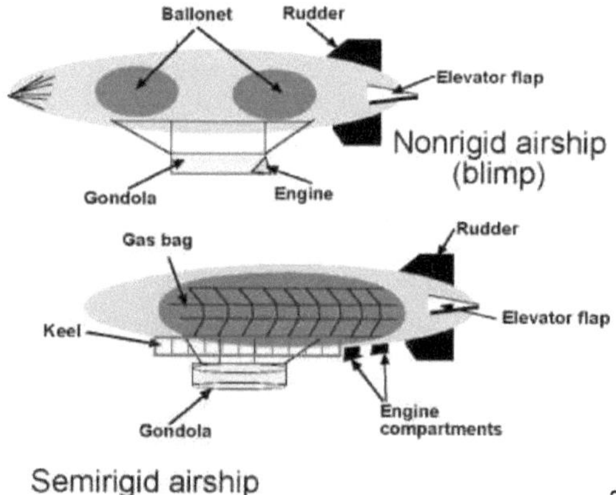

---

[17] http://www.galaxy.bedfordshire.gov.uk/webingres/bedfordshire/vlib/0.digitised_resources/0.images/r101_mast.jpg

[18] http://www.century-of-flight.net/Aviation%20history/up%20to%20WW%201/Dirigibles2.htm

[19] For more details about "*The Italia*" visit: http://90north.tripod.com/italiacrash1.htm

[20] http://z.about.com/d/inventors/1/0/6/U/airship_types1.jpg

Because non-rigid airships often collapsed under stress, a fixed keel - similar to that of a boat - was added under the envelope to give it strength or enclosed the gas bag inside a frame.[21] The keel acted as a sort of metal spine which helped maintaining the envelope's shape and supported the gondola and engines. The ballonets were needed to keep the envelope pressure constant. These semirigid airships were often used for reconnaissance flights.

3.3 Non-rigid Airships

Non-rigid airships, also known as Blimps, are the most common form nowadays. Basically they are large, dirigible, inflated gas bags, more like a hot air balloon with no interior skeleton or supporting structure. Their shape is maintained by an internal overpressure. The only solid parts are the passenger capsule and the tail fins. These non-rigid airships use ballonets, which are air bags located inside the outer envelope that expanded or contracted to compensate for changes in the gas. The airship-homepage *http://www.airshipman.com* describes the use of modern airships:

"No other advertising vehicle turns more heads or holds greater intrigue than airships. They are used as both a unique alternative for advertising and promotional purposes going beyond common media venues to give your company maximum exposure."[22]

Nearly all the airships currently flying for publicity use are of the type of non-rigid airship, e.g. the Fuji Blimp in New York[23] as well as the three well-recognized blimps of Goodyear[24] in the USA. But where does the term "Blimp" come from? According to research done by the Goodyear Company, the term "Blimp" originates from Lieutenant "Cunningham of Great Britain's Royal Navy Air Service. He commanded the air station at Capel, England, during World War I. As the story goes, while conducting a weekly inspection of the station, Cunningham playfully flipped his thumb at the envelope of His Majesty's Airship SS-12 and an odd noise echoed off the taut fabric. "Blimp!" he cried out humorously, imitating the sound. As they say the rest is history."[25]

---

[21] http://inventors.about.com/od/astartinventions/ss/airship_9.htm

[22] http://www.airshipman.com

[23] For pictures and more details: http://www.airshipman.com/NewsArticles/News30aug04.htm

[24] For pictures and more details: http://www.goodyearblimp.com

[25] http://www.goodyearblimp.com/history/index.html

4. Hot Air Airships

Hot air airships, also known as thermal airships, are counted as a fourth kind even though they are technically part of the non-rigid category. Hot air airships are derived from traditional hot air balloons. Early models were almost like elongated balloons with an engine and tail fins added. Pretty soon envelopes were lengthened and the tail-fins and rudder were pressurized by air from the wash of the propeller. Newer hot air airships maintain their shape with internal overpressure in the whole envelope, a feature which older models did not have. Today, their main uses are in sportive competition, advertising and environmental work.[26]

5. Aeroscraft

The above mentioned airships are either outdated (like the rigid- and the semi-rigid-airships) or mainly used for advertising like nearly all of the non-rigid airships. During the last decade several companies have re-discovered the lighter-than-air concept and have been introducing safer, more efficient airships or are about to develop new concepts for use in the near future. On of these concepts is presented by *Worldwide Aeros Corporation*[27], the so-called *Aeroscraft*, which is described as "a new paradigm in air transport […because it…] utilizes adjustable buoyant and dynamic lift, creating unique capabilities beyond what is available from any other air platform today."[28] With about 14 million cubic feet of helium inside the aerodynamic shape and additionally six turbofan jet engines, the remaining lift is provided to keep the Aeroscraft at its cruising altitude of around 8,000 feet (ca. 2,700 meters). It is designed as a luxury cruise ship that sails through the air (with up to 250 passengers) with the capability of hovering and vertical takeoff and landing, making it independent from airports.[29] That's why it has potential as a cargo ship too, being able to transport up to 500 tons of even huge items of cargo without disrupting road traffic and even faster than trucks.[30] Right now the Aeroscraft is in the manufacturing process and a prototype should be finished by 2010.

---

[26] For more details: http://www.hotairships.com

[27] Company's Homepage: http://www.aerosml.com

[28] http://www.aeroscraft.com

[29] http://www.aerosml.com/ML866/AEROS_ML866.pdf

[30] http://science.howstuffworks.com/aeroscraft1.htm Worldwide Aeros Corp. Aeros-D4 and Aeros-D8.

6. New developments

Since the price of fuel rises and the concern for the environment grows, entrepreneurs are taking another look at the usage of airships today and a new generation of dirigibles is being considered by governments and private companies as well. The development of new materials and new means of propulsion have paved the way for this new dirigibles. In Germany, the *Deutsche Zeppelin-Reederei* was founded in 2001 as a successor to the operator of *The Hindenburg* to revive the operation of Zeppelin airships in the sectors tourism, advertisement and special scientific missions. In April 2001, the new *Zeppelin NT 07*[31] received its type certification of the German Federal Office of Civil Aeronautics and sightseeing flights over Lake Constance were offered on a daily basis. They expanded their activities and during the first 5 years of passenger operations, the famous Zeppelin has transported more than 55,000 passengers.[32] Today they are operating three Zeppelin airships (in Japan, South-Africa and Germany) and are offering 12 different sightseeing tours over southern Germany on a duration between 30 and 120 minutes.[33]

To get an impression of the enormous size of airships, the new Zeppelin NT is compared with its successors as well as a Boing 747-400 aircraft:

GIFFARD AIRSHIP (MAIDEN FLIGHT: 1852)
Created by Henri Giffard, a French engineer and architect. The first powered aircraft, it used a propeller driven by a lightweight steam engine to travel 17 miles on its only flight.

144 feet long; 1 pilot

Behemoths in the Air

Dirigibles, which use enormous balloonlike cells filled with lighter-than-air gas for lift, are drawing renewed interest.

LZ-129 HINDENBURG (1936)
More than 1,000 passengers traveled on 18 trans-Atlantic trips to Rio de Janeiro and Lakehurst, N.J. But in May, 1937, the Hindenburg caught fire and crashed while landing in Lakehurst. It was filled with hydrogren, a buoyant but highly inflammable gas, not helium, which is inert.

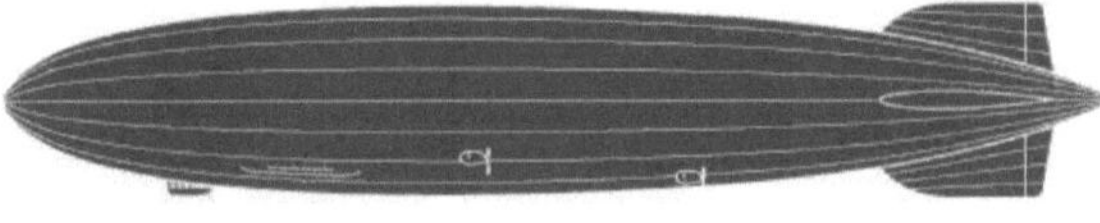
803 feet long; about 50 passengers and about 50 crew

GOODYEAR GZ-20 (1968)
Well-known due to their prominence at sporting events, these blimps are some of the current airships used for advertising and broadcast.

192 feet long; 6 passengers and 1 pilot

BOEING 747-400
231 feet long; 400 passengers

ZEPPELIN NT (1997)
Built by ZLT Zeppelin Luftschifftechnik, the fourth NT airship is expected to begin service in California this fall, for sightseeing tours.

246 feet long; 12 passengers and 2 crew

Sources: Boeing, Deutsche Zeppelin-Reederei; Goodyear Blimp; "Graf Zeppelin & Hindenburg" by Harold G. Dick and Douglas H. Robinson, 1985 (Smithsonian Institution Press); "The Airships: A History" by Basil Collier, 1974 (G. P. Putnam's Sons)

THE NEW YORK TIMES **34**

---

[31] Zeppelin NT 07 with NT standing for New Technology.

[32] http://www.zeppelinflug.de/seiten/E/zeppnt_histo.htm

[33] http://www.zeppelinflug.de/seiten/E/default.htm

[34] http://www.nytimes.com/imagepages/2008/07/05/business/05dirigibles.graphix.ready.html

But tourism is not the only use envisioned for such airships: France's postal service, La Poste, seems to have some of the most ambitious plans in mind: using airships on routes between France and Corsica or the Antilles in an effort to reduce emissions.[35] And in the United States, research into dirigibles continues, but mainly for military purposes.[36] In Germany, a huge project of developing an airship for transport of cargo was planed around 10 years ago. The company *CargoLifter AG*, based in Berlin, developed a "260-meter-long, 65-meter-wide airship for the transport of oversized and heavy goods weighing up to 160 metric tons: the CargoLifter CL 160. This "flying crane" will be able to carry expensive heavy transports faster, more simply and more cost effectively without ever touching the ground while remaining almost entirely independent from the local infrastructure. In a second product-line, which will be ready before the CL 160, the CargoLifter CL 75 AirCrane transportation balloon with a diameter of 61 meters will be able to carry loads of up to 75 tons."[37] But this ambitious project failed due to financing problems, the *CL 160* was never built and the company filed for bankruptcy just two years after going public.[38] The immaterial goods of the closed *CargoLifter AG* were purchased by the *ZLT Zeppelin Luftschifftechnik GmbH & Co KG / Deutsche Zeppelin-Reederei GmbH*, which is continuing conduct research in airship technology.[39]

7. The pros and cons

The advantage of light-than-air vehicles over heavier-than-air crafts is that static lift sufficient for flight is generated by the lifting gas (modern airships use dynamic helium volume) and require no engine power. If engines are added to enable maneuverability, the airship gets additional thrust and with modern concepts (e.g. for high altitude airships) photovoltaic cells are being added to reduce the need to land to refuel, thus they can remain in the air until consumables expire.

---

[35] http://www.nytimes.com/slideshow/2008/07/06/business/0706-BLIMP_3.html

[36] http://www.defenseindustrydaily.com/walrus-heavylift-blimp-getting-off-the-ground-01103

[37] http://www.presseportal.ch/de/meldung/100010897

[38] http://www.manager-magazin.de/geld/artikel/0,2828,198297,00.html

[39] http://www.zeppelinflug.de/seiten/E/faqs.htm

Another potential advantage of airships over fixed wing aircraft is cargo capacity, which is the predominant field of actual research studies and further development. Some airships being considered as CargoLifters or SkyFreighters could carry up to 1,000 tons of cargo up to 12,000 miles, making it interesting especially for military operations.[40] Thus, the cargo capacity of just one airship is roughly equivalent to the payload of 13 C-17 aircrafts. Because it is able to land vertically on both land and water, and is able to hover, airships can potentially deliver their cargos everywhere. What makes it interesting for the military is that airships may be more survivable than they appear, because despite their large sizes, airships have smaller radar and infrared signatures than normal transport and cargo aircraft. But on the other side their lower cruising level makes it more vulnerable by enemy's weapons.

The disadvantages are that airships have a very large reference area making them sensitive to wind & weather and are only operable under visual meteorological conditions[41]. They have a comparatively large drag coefficient, thus a larger drag force compared to that of airplanes and even helicopters. Because of its large flat plate area and wetted surface, a practical speed limit is reached around 80–100 knots (130–160 km/h), which is much slower than cargo-airplanes flying around 450 knots. Thus airships are used where speed is not critical - but on the other side the very large payload is unmatchable and makes up for the slower speed.

8. Summary

To feel the fundamental difference between the described lighter-than-air-systems and fixed wing aircraft, just take a ride in a hot-air balloon and feel how lift is generated by the buoyancy of hot air as the lifting gas. Riding a balloon is fun and maybe the most successful lighter-than-air concept in the world.

Airships are able to hover or to fly at low speeds but require a huge hull volume which, in turn, causes high friction drag and limits the maximum speed. With the Hindenburg-disaster in 1937, the era of these airships ended abrupt and the time of

---

40 http://www.defenseindustrydaily.com/walrus-heavylift-blimp-getting-off-the-ground-01103

41 Definition of Visual Meteorological Conditions: http://www.skybrary.aero/index.php/VMC

rigid- and semi-rigid airships was definitively over. Since around one decade, new dirigibles are invented again and with the Zeppelin NT they are presently established as a sightseeing attraction in a safe and successful way.

With soaring fuel costs and greenhouse gas concerns, non-rigid airships like the zeppelins might get their second chance to be a relevant mode of transportation. Nevertheless, they are expected by most aviation experts to remain niche vessels for ferrying tourists, advertising and occasional scientific payloads. If cost-extensive projects like the CargoLifter or the Aeroscraft will ever be able to operate in a (cost) effective manner remains to be seen.

Airship *Zeppelin LZ New Technology 07*
filled with incombustible helium.

---

42 http://www.zeppelinflug.de/fotogalerie/bilder.htm

9. Sources

<u>Internet Sources: Documents</u>

The New York Times: ROMA was largest semi-rigid airship, New York, 22.02.1922.

http://query.nytimes.com/mem/archive-free/pdf?res=9B0CEFD71E30EE3ABC4A51DFB4668389639EDE

Worldwide Aeros Corp.: Aeroscraft Model Brochure.

http://www.aerosml.com/ML866/AEROS_ML866.pdf

<u>Internet Sources: Homepage informations</u>

Calculation of air density

http://www.peacesoftware.de/einigewerte/luft.html

Goodyear Company - Fleet of Blimps

http://www.goodyearblimp.com

Information about Aeroscraft

http://www.aeroscraft.com

Information about Airships

http://www.airships.net/lz127-graf-zeppelin/design-technology

http://www.airshipman.com

http://www.globalsecurity.org/military/systems/aircraft/airship.htm

http://90north.tripod.com/italiacrash1.htm

http://www.infovisual.info/05/093_en.html

http://inventors.about.com/od/astartinventions/ss/airship_9.htm

http://www.galaxy.bedfordshire.gov.uk/webingres/bedfordshire/vlib/0.digitised_resources/0.images/r101_mast.jpg

Information about CargoLifter AG

http://sebastianrusche.com/cargolifter

http://www.presseportal.ch/de/meldung/100010897/

http://www.manager-magazin.de/geld/artikel/0,2828,198297,00.html

Information about Zeppelin

http://www.altfrankfurt.com/Spezial/Zeppelin/GrafZeppelin

http://www.century-of-flight.net/Aviation%20history/up%20to%20WW%201/Dirigibles2.htm

http://www.century-of-flight.net/Aviation%20history/coming%20of%20age/zeppelin.htm

http://www.zeppelinflug.de/seiten/E/zeppnt_histo.htm

http://www.zeppelinflug.de/fotogalerie/bilder.htm

http://www.zeppelinflug.de/seiten/E/faqs.htm

Information about hot-air balloon

    http://www.eballoon.org/balloon/how-it-works.html

    http://www.eballoon.org/balloon/commercial-leisure.html

    http://science.howstuffworks.com/hot-air-balloon1.htm

    http://blog.adventureballoons.co.uk/category/the-balloon-envelope

    http://www.stevefossett.com/html/press_archives/balloon_04july02.html

Information about hot-air airships

    http://www.hotairships.com

    http://www.airshipinitiatives.com

Information about U.S. Walrus Program

    http://www.defenseindustrydaily.com/walrus-heavylift-blimp-getting-off-the-ground-01103

Information about weather conditions for flights

    http://www.skybrary.aero/index.php/VMC

The New York Times Article "Why fly when you can float", published July 5, 2008.

    http://www.nytimes.com/2008/07/05/business/worldbusiness/05dirigible.html?_r=1

The New York Times Article "A New Age for Dirigibles", published July 6, 2008.

    http://www.nytimes.com/slideshow/2008/07/06/business/0706-BLIMP_5.html

All internet-based references were effective and available as of 9. November 2009.

YOUR KNOWLEDGE HAS VALUE

- We will publish your bachelor's and
  master's thesis, essays and papers

- Your own eBook and book -
  sold worldwide in all relevant shops

- Earn money with each sale

Upload your text at www.GRIN.com
and publish for free